# FLORA OF TROPICAL EAST AFRICA

## BIXACEAE

### D. M. Bridson

Trees or shrubs. Leaves spiral; blades simple, entire, pendulous to perpendicularly hinged from the long erect petioles; petioles with a swelling (pulvinus) at the apex and base; stipules paired, very quickly caducous, leaving a horseshoe-shaped, scale-like scar with a sessile gland on either side of the petiole-base. Flowers hermaphrodite, regular, in terminal corymbs or panicles. Receptacle with sessile glands around the rim. Sepals (4–)5, free, imbricate in bud, caducous on expansion of the flower. Petals (4–)5(–7), free, imbricate in bud. Stamens numerous, inserted on the annular disc; filaments free; anthers with 2 horseshoe-shaped thecae, dehiscing by short slits on the apical bend. Ovary superior, 1-locular; ovules numerous on 2(–3) parietal placentas; style simple, elongated; stigma shortly 2-lobed. Fruit an echinate, setose or smooth capsule, loculicidally bivalved (rarely 3-valved). Seeds numerous; testa fleshy; endosperm copious; embryo large.

A family comprising a single genus native to Central and South America. Of the four species one is commonly cultivated throughout the tropics.

## BIXA

L., Sp. Pl. 1: 512 (1753) & Gen. Pl., ed. 5: 228 (1754)

Characters of the family.

**B. orellana** *L.*, Sp. Pl.: 512 (1753); Oliv., F.T.A. 1: 114 (1868); P.O.A. B: 404 (1895); Sim., For. Fl. Port. E. Afr.: 11, t. 2/A (1909); Volkens in N.B.G.B. App. 22, 3: 68, 110–111, fig. 57 (1910); V.E. 3(2): 540, fig. 243 (1921); Pilger in E. & P. Pf., ed. 2, 21: 314, fig. 139 (1925); C.F.A. 1: 77 (1937); T.T.C.L.: 74 (1949); Backer & Heemstede in Fl. Males., ser. 1, 4: 239, figs. 1, 2 (1951); Keay in F.W.T.A., ed. 2, 1: 183 (1954); E.P.A.: 594 (1959); Wild in F.Z. 1: 261 (1960); F.F.N.R.: 263 (1962); Hutch., Fam. Fl. Pl., ed. 3: 253, fig. 74 (1973); Villiers in Fl. Gabon 22: 59, t. 15 (1973). Type: Brazil and Mexico, sterile specimen in Clifford Herbarium (BM, syn.!)

Shrub or small tree 1–9(–15) m. tall; young branches densely covered with rust-coloured peltate scales, becoming glabrescent with age. Leaves petiolate; blades ovate*, 5–25 cm. long, 3·3–16·5 cm. wide, apex long acuminate, base cordate or less often truncate (very rarely obtuse), very young leaves with rust-coloured sessile scales beneath, soon becoming glabrescent or glabrous, with 5–6 pairs of lateral nerves, the second pair always arising from the base together with the first pair, prominent

---

\* One specimen, *Alston* 2109, cultivated in Ceylon, has leaves which are very deeply 3(–5)-lobed. The only intermediate seen, *Mathews* 3100 from Peru, has a ± 5 undulate margin.

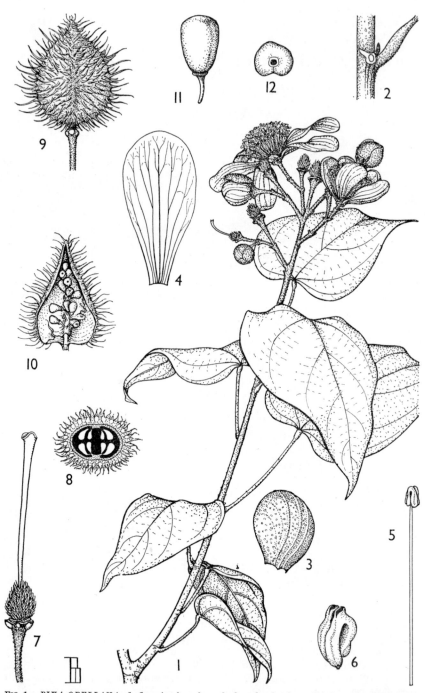

FIG. 1. *BIXA ORELLANA*—1, flowering branch, × ⅔; 2, node, showing petiole-base and stipular scar, × 2; 3, sepal, × 2; 4, petal, × 2; 5, stamen, × 4; 6, anther, dehisced, × 14; 7, pistil, × 3; 8, transverse section of ovary, × 6; 9, fruit, × ⅔; 10, valve of dehisced fruit, × ⅔; 11, 12, seed, viewed from side and apex respectively, × 3. 1, from *Tanner* 2315; 2, 9, from *Latilo & Daramola* in *FHI*. 28831; 3–8, from *Angus* 3060; 10–12, from *Culwick* 3.